ICE CAPS & GLACIERS

Design	David West
	Children's Book Design
Designer	Keith Newell
Editorial Planning	Clark Robinson Limited
Editor	Yvonne Ibazebo
Picture Researcher	Emma Krikler
Illustrators	Mike Saunders
	and Simon Tegg
Consultant	David Flint
	University lecturer

© Aladdin Books Ltd 1992

First published in
the United States in 1993 by
Gloucester Press
95 Madison Avenue
New York, NY 10016

Library of Congress Cataloging-in-Publication Data

Twist, Clint.
    Ice caps and glaciers / Clint Twist.
        p.   cm. — (Hands on science)
    Includes index.
    Summary: Describes what ice from the polar regions can tell us
about past climatic conditions and our changing atmosphere and
examines the role of water in its solid state within the environment.
Including projects and other hands-on activities.
    ISBN 0-531-17396-8
    1. Ice—Juvenile literature. [1. Ice. 2. Polar regions.] I. Title.
II. Series.
GB2403.8.T88   1993
551.3'1—dc20   92-33917   CIP   AC

Printed in Belgium

HANDS · ON · SCIENCE

ICE CAPS & GLACIERS

CLINT TWIST

GLOUCESTER PRESS
New York · London · Toronto · Sydney

CONTENTS

This book is all about ice caps and glaciers. The topics covered range from the vast ice sheets at the Poles to the moving valley glaciers of high mountain regions. The book talks about the many landforms left behind by glaciation, such as the beautiful fjords along the coast of Norway, or the kames and eskers created by glacial deposition. The book also talks about the ways in which glaciers help scientists find out about the state of the earth's climate millions of years ago. There are "hands on" projects for you to do which use everyday equipment. There are also "did you know" panels of information for added interest.

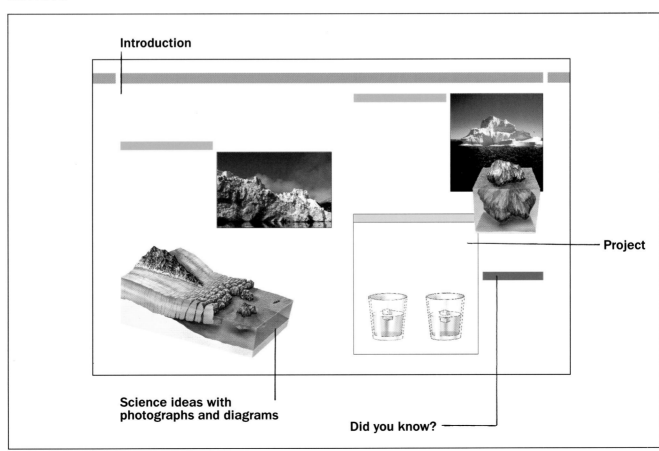

Introduction

Project

Science ideas with photographs and diagrams

Did you know?

Although the earth's climate is fairly warm, some parts, such as the Poles and high mountain ranges, remain covered with glacier ice all year round. Glaciers vary greatly in size and depth. Some of them are quite tiny and measure only a few hundred feet long. Others cover vast areas of land and may be thousands of miles deep. Glaciers rarely remain still. Their massive weight causes the ice to flow slowly downhill, scraping the rock beneath and creating new landforms.

   Less than 10,000 years ago, the earth's climate was much colder than it is today. Glaciers extended over much of North America, Europe, and Asia, and huge sheets of ice covered Greenland and Antarctica. Even in warmer parts of the world, such as Africa, South America, and New Zealand, glaciers formed in the high mountain areas. These glaciers gradually formed many different geographical features, such as the Great Lakes and valleys of North America.

Glaciers are an important source of water and electricity.

THE COLD WORLD

Some parts of the world are so cold that they are covered with permanent layers of ice. The largest layers occur around the North and South poles, where they spread out to form ice caps thousands of feet thick. Smaller glaciers are found in mountain valleys all over the world. These are known as valley glaciers.

EXTENT TODAY

Glacier ice covers about 6.5 million square miles of the earth's surface, just over ten percent of the total land area. Nearly all of this ice (98 percent) occurs at the Poles. Valley glaciers account for the remaining two percent.

Huge ice sheets have advanced across Europe and North America during successive glacial periods, and at one time the whole of Great Britain was covered with a large mass of ice.

Today, glaciers can be found in areas that have very cold winters and short, cool summers, such as Greenland and Iceland. Glaciers can also be found in mountains that have very high altitudes, such as Mount Kilimanjaro in Tanzania, and in high mountain chains like the Alps, Andes, Rockies, and Himalayas.

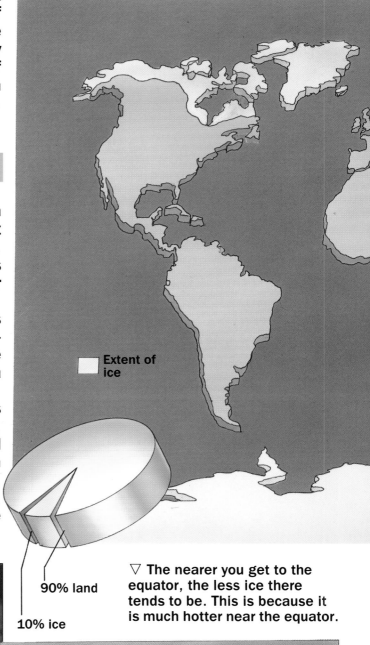

Extent of ice

90% land

10% ice

▽ The nearer you get to the equator, the less ice there tends to be. This is because it is much hotter near the equator.

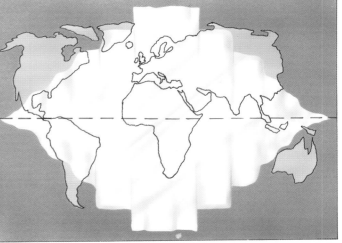

A view of a glacier and icebergs.

THE POLES

The Poles are very cold because of their geographical position. Situated at the very top and bottom of the globe, the Poles receive sunlight for only half of the year. The remaining six months are spent in complete darkness. Even at the height of the polar summer, the sun never reaches its full height in the sky. As a result, it shines at an angle on the Poles, and the amount of heat reaching the land is very much reduced.

Glaciers are made up of snow, and at the Poles, snow falls throughout the year. This snow does not melt but becomes compressed into glacial ice. The ice and snow that collect have an extra cooling effect, for the white surface reflects a high proportion of the sunlight and absorbs little heat.

The region around the North Pole is known as the Arctic, and the ice covering in this area extends southward over most of Greenland. Around the South Pole, ice completely covers the continent of Antarctica, making it the coldest and iciest region in the world.

△ Today, ice covers ten percent of the earth's land surface. Nearly all of this ice can be found at the Poles.

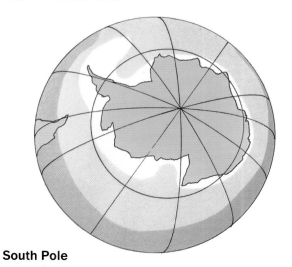

South Pole

Winter ice limit

Summer ice limit

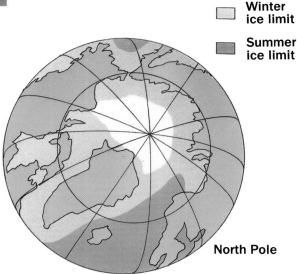

North Pole

△ About 99 percent of the world's fresh water is locked into the ice sheets of Antarctica and Greenland.

In the past, the earth's climate was much colder than it is today. The polar glaciers extended much further south and north, and vast areas of land were buried under a half-mile-thick sheet of ice. These periods are known as the Ice Ages, and the most recent one ended only about 10,000 years ago.

EXTENT OF THE ICE

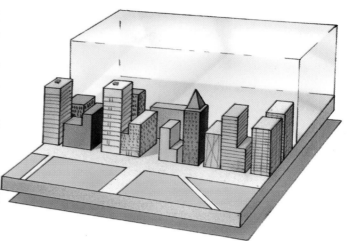

△ The illustration above shows the size of a glacier against the high-rise buildings of New York City.

Between 400,000 and 10,000 years ago, our planet experienced a series of Ice Ages separated by warm periods. We are probably living in one of these warm periods, called interglacials, today. During the last Ice Age, the glaciers reached their furthest extent and covered about 30 percent of the earth's land surface.

The effects of the Ice Age were greatest in the Northern Hemisphere, where huge sheets of ice covered much of North America, Europe, and parts of Asia. South of the equator, the glaciers covered about one-third of South America, and parts of Australia. One major effect of the last Ice Age was a fall in sea levels. Millions of tons of water were used in forming glaciers, and in many parts of the world, land that had been submerged under water became exposed. The Black Sea was turned into a lake, and around the Pacific, New Guinea was connected to Australia. As the glaciers melted, sea levels rose and the landscape took the form it has today.

At their greatest extent, glaciers covered about 30 percent of the earth's land surface.

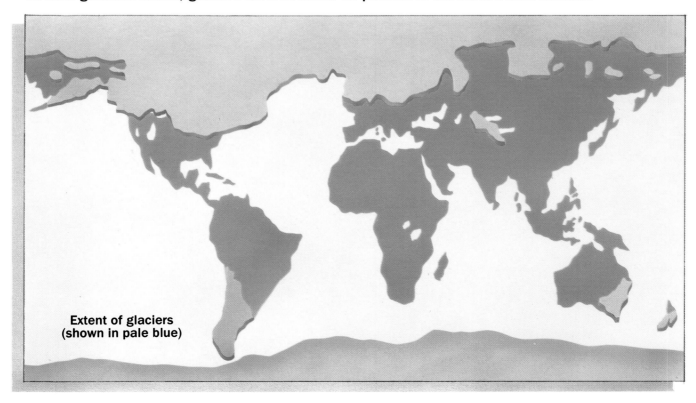

Extent of glaciers
(shown in pale blue)

EFFECTS ON LAND

During the last Ice Age, the amount of habitable land on earth was greatly reduced. Animals, including our ancestors, were forced to move away from the advancing glaciers toward the equator. With so much water turned into ice, the climate was much drier than it is today. The rain forests shrank so much that they almost disappeared completely.

Further north and south of the equator, the effects of the glaciers were even more dramatic. The large masses of thick ice hid the features of the land beneath. Glaciers also helped to form new landscapes, as they are among the most powerful land-shaping forces on earth. North America and northern Europe are full of features carved out by glaciers. When the glaciers melted, the flood of water released added new features, such as lakes and valleys, to the landscape.

Death Valley in California.

ICE TO WATER

The effect of melting glaciers does not alter sea levels. You can demonstrate this for yourself. Fill a margarine tub with water and place it in a freezer until the water has completely turned to ice. Remove the ice block from the tub and place it in a shallow bowl. Add water until the ice block floats. Measure the level of water in the bowl with a ruler. Wait until the ice has melted and measure the depth again. The "sea level" in the bowl should not have risen.

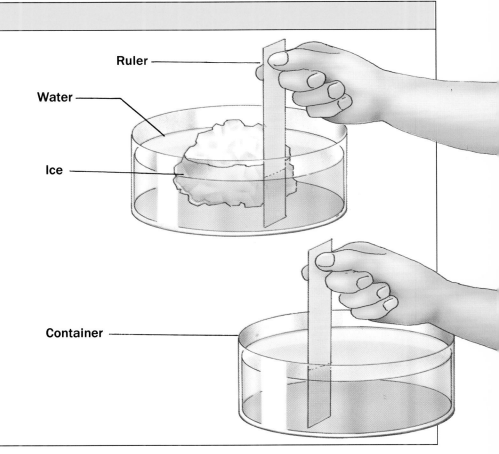

Altogether, the polar regions around the world contain several million cubic miles of ice. This represents almost all of the fresh water found on our planet. Most of the ice at the Poles lies on solid rock. Floating ice in the sea makes up less than one percent of the total volume of ice at the Poles.

ANTARCTIC

In the continent of Antarctica, ice covers over 4.6 million square miles, and in some places it is more than 13,000 feet thick. Beneath the ice lies a varied landscape of mountain ranges, lowlands, and valleys. The massive weight of the ice has depressed the earth's crust, and many parts of Antarctica have been pushed down below sea level. In places where these depressions open out into the sea, ice shelves have formed. An ice shelf is a thick layer of ice that starts on land and extends out into the sea. Apart from a few plants and insects, Antarctica is completely uninhabited. There are also 30 science and weather stations on the continent.

△ The Arctic is comparatively warm. At the North Pole, the average midsummer temperature is 32°F. On Greenland, temperatures may drop to 5°F.

ARCTIC

In contrast, much of the Arctic ice consists of an island of pack ice floating in the middle of the Arctic Ocean. The ice has an average depth of seven feet, but in some places it is considerably thicker. Even so, there is clear water under the North Pole, and submarines are able to travel beneath the ice.

On Greenland, however, a thick layer of ice extends for nearly one million square miles, covering almost all of the land. Only the narrow strips along parts of the coast are free from ice, because these areas are washed by warm ocean currents. Most of the people of Greenland live in the southwest where summer temperatures average 40°F.

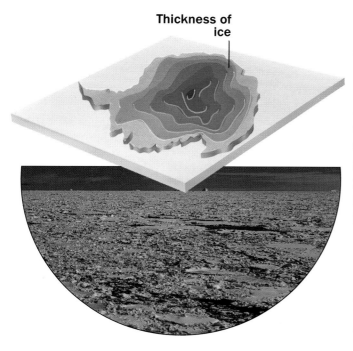

Thickness of ice

◁ Antarctica is the coldest place in the world. The midsummer temperatures rarely rise above −22°F. In winter, temperatures fall below −90°F.

ICE CAPS

The term ice cap is usually used to describe ice that can be found at the Poles, because they "cap" the top and bottom of our planet. But the ice on Antarctica and Greenland is most often called ice sheets.

An ice cap is an extensive glacier that covers less than 20,000 square miles. The term ice sheet is used for larger masses of ice. A large area of permanent pack ice is known as an ice field, because of its flat surface and uniform thickness.

Glaciers are frequently thought of as frozen rivers. Despite being frozen and compressed into dense ice, they still tend to flow downhill. Within the vast polar ice sheets, great rivers of ice slowly flow over the underlying landscape toward the sea. These rivers of ice are called continental glaciers.

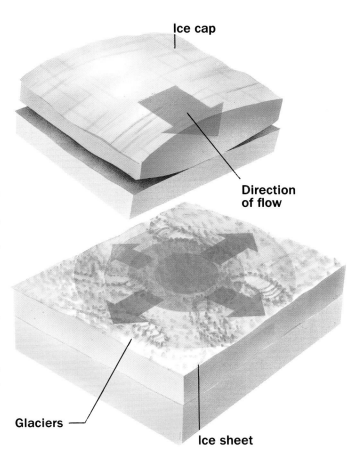

Ice cap

Direction of flow

Glaciers

Ice sheet

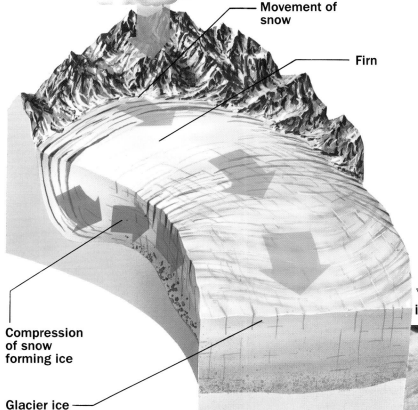

Snowfall

Movement of snow

Firn

Compression of snow forming ice

Glacier ice

△ Glaciers are given different names, depending on how large they are. Ice caps and ice sheets can cover thousands of square miles of land.

◁ All glacier ice comes from snow. Snowflakes melt very slowly in the intense cold around the Poles. It takes about one year for the snow crystals to turn into the hard, dense crystals known as firn. As more snow falls, the firn begins to turn into larger crystals of ice.

▽ A cross section through an ice core shows layers of snowfall.

The ice sheets at the Poles receive about 12 inches of fresh snow every year. Combined with the force of gravity, the weight of the accumulated snow drives continental glaciers toward the sea. Most of the water lost from the polar ice sheets occurs when masses of ice (icebergs) break off the sheet and float out to sea.

JOURNEY TO THE SEA

Buried beneath the ice sheets are mountain ranges that are as high as the European Alps. The largest glaciers are found in Antarctica. The Lambert Glacier is more than 40 miles wide, and over 250 miles long.

Where the amount of snow falling on the glacier is the same as that lost through icebergs, the glacier remains almost stationary. But when the snowfall is more, the glacier moves forward a few feet a day. One example was the Hubbard glacier in North America, which advanced at around 32 feet a day in 1986. The fastest-moving glacier is the Quarayaq in Greenland, which moves about 72 feet a day.

In Greenland, glaciers flow straight into the sea and begin to melt. Cracks appear in the glacier, and large ice blocks form. These break away to form icebergs which can be found in the North Atlantic. This process is known as calving.

Ice cliffs at the edge of Svitjobreen glacier in Spitsbergen.

◁ A continental glacier flows toward the sea. As the ice melts, bits of it break off to form icebergs. This process is known as calving.

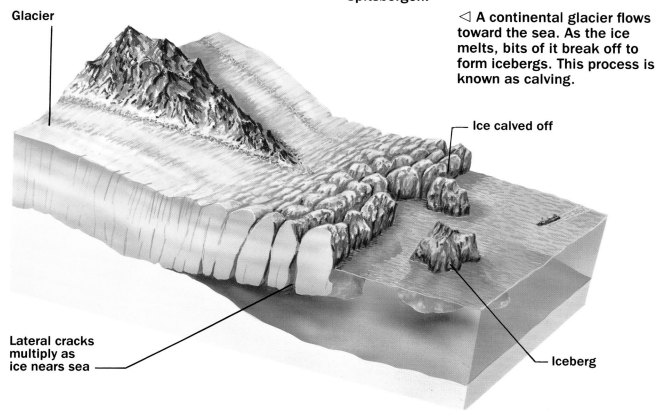

Glacier

Ice calved off

Lateral cracks multiply as ice nears sea

Iceberg

BERGS

In the Northern Hemisphere, icebergs tend to be relatively small. Although some measure a few miles, the vast majority are only a few inches long. Very small icebergs are called growlers. Smaller lumps of ice are known as "bergy bits." Northern icebergs usually have a very irregular shape, and even the largest melt within a year or two.

Around the edges of the Antarctic ice sheet, huge tabular icebergs can be found drifting in large numbers in the ocean. Icebergs more than 12 miles long are common. These massive ice-islands remain intact for many years, and only begin to melt when they eventually drift toward the equator.

Sea ——

Iceberg

△ An iceberg floats with 80 percent of its bulk underwater. There is usually a groove on them at sea level caused by wave action.

HOW MUCH BELOW

Ice floats at different depths in fresh water and salt water. You can measure the difference for yourself with an ice cube in a glass. Dissolve some salt in a glass of water. See how much of the ice cube sticks out above the surface of the glass. Now use fresh water and compare the results.

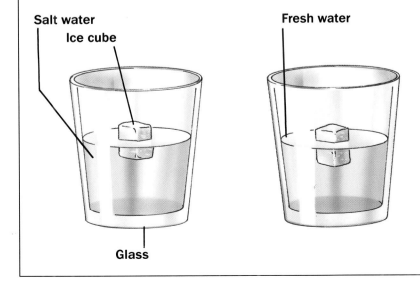

Salt water

Ice cube

Fresh water

Glass

DID YOU KNOW

The largest icebergs produced in the continent of Antarctica measure more than 60 miles (100 km) in length. Three tabular bergs that calved from the Filchner ice shelf in 1986 had a combined area of more than 4,000 square miles. Each of them was bigger than the country of Luxembourg, and had ice more than 1,600 feet thick.

The formation of glaciers and ice sheets requires abundant snowfall. Around the fringes of the Arctic is an area where it is too dry and windswept for much snow to fall. The main feature of this tundra climate is frost, and the ground is always frozen into permafrost hundreds of feet deep.

PERMAFROST

The permafrost zone stretches across Alaska, northern Canada, Lapland, and northern Russia. Beneath the surface, the ground is frozen solid and in parts of Siberia, the permafrost extends to a depth of 1,600 feet. During the spring and summer months, only the top few inches of ice ever thaw out completely, and then only for the brief period before the autumn freeze.

One of the main effects of this freeze-thaw cycle is to flatten out the landscape. In winter, the surface layer expands as it freezes, raising the soil. During spring, the thawing process makes the surface layer muddy, and it slides down the gentlest slope. This allows the soil to settle more evenly. Tundra soils are usually well graduated, with the smallest particles at the surface.

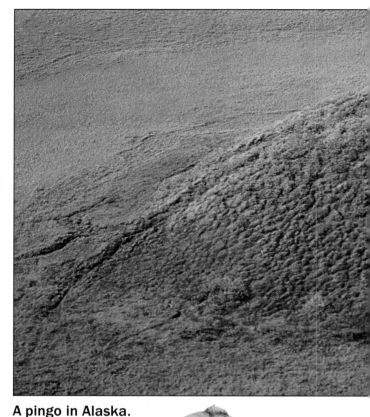
A pingo in Alaska.

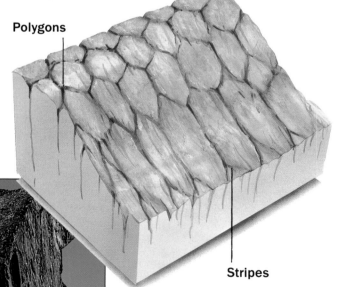
Polygons

Stripes

△ On level surfaces, frost action often shapes the ground into frozen polygons with almost regular sides. On slopes, the sides extend into stripes.

◁ A large rock near the summit of Bella Tolla in Switzerland. The ground here always remains frozen.

ICE ACTION

The freeze-thaw cycle of ice has a destructive effect on rocks exposed at the surface. Water that seeps into cracks in the rocks expands as it freezes, shattering the rocks. A similar process extends all the way into the permafrost.

During the winter freeze, narrow cracks appear in the soil surface. In spring, water seeps into these cracks and as the water freezes, the cracks widen and deepen. In time, these cracks develop into subsurface ice wedges up to 50 feet deep and 10 feet across. When the ice wedge finally melts, the resulting depression it leaves may slowly fill up with fine sediment.

The most dramatic feature of permafrost landscapes are pingos, small conical mounds which may be up to 330 feet high. A pingo is formed when underground water accumulates just below the surface. As the water freezes, it pushes the surface up into a dome shape. When the water melts, the enlarged cavity accumulates even more water. Eventually the ice breaks through the surface and melts. Then the soil covering collapses and produces a crater shape.

▽ This stone in northeast Greenland has been shattered by frost.

Ice

Gas pressure

△ Below ground level, expansion due to freezing can push soil up to the surface of a glacier.

A valley glacier is a long, narrow mass of ice that fills high mountain valleys. In total, there are about 10,000 valley glaciers scattered throughout the world, from the Andes in South America to the European Alps. Altogether, valley glaciers account for about two percent of the world's ice.

FORMATION

Whatever their size, all valley glaciers form in the same way — from a small hollow high up in the mountain filled with snow. Over the years, freeze-thaw action enlarges the hollow, and the snow turns into firn. As more snow accumulates, the firn turns into glacier ice. At this stage, the hollow is known as a cirque. Eventually the glacier overflows the lip of the cirque and begins to slide downhill.

Although it flows much slower than liquid water (on average only about 40 feet a day), a valley glacier behaves rather like a river. Where one glacier runs into another, they converge just like rivers do. Many valley glaciers are hundreds of miles long. For example, the Bering Glacier in southern Alaska is 125 miles long.

Not all the ice within a valley glacier moves at the same speed. Ice in the center of a valley glacier moves quite fast. But at the edge of the glacier, the ice is slowed down by friction against the sides of the mountain. Also, the underlying bedrock beneath the glacier tends to slow the moving ice down.

Hardangervidda glacier in Norway

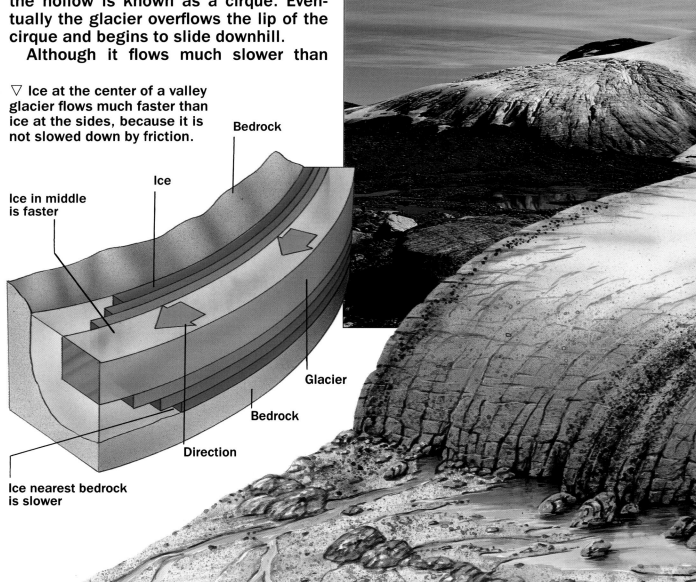

▽ Ice at the center of a valley glacier flows much faster than ice at the sides, because it is not slowed down by friction.

Bedrock

Ice

Ice in middle is faster

Glacier

Bedrock

Direction

Ice nearest bedrock is slower

EROSION

Glaciers are extremely powerful agents of erosion. As they creep along a valley, they change the shape of the underlying rock. Glaciers erode the landscape in two ways — by plucking and by abrasion.

At the sides and bottom of a valley glacier, friction melts some of the ice, and the water seeps into cracks in the underlying rock. As the water freezes, the glacier plucks fragments of rock away as it slides downhill.

Rock fragments become incorporated into the ice, and get dragged along with the glacier. The rocks make the glacier act like an enormous file. As it moves downhill with the rocks, the glacier cuts and shapes the rock beneath.

▽ A valley glacier accumulates snow at the top of a mountain. As it flows downhill, it erodes the landscape by plucking and abrasion.

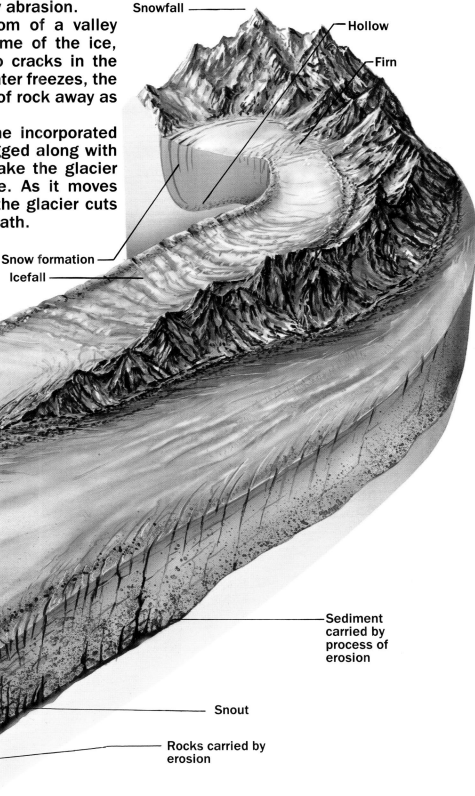

Snowfall

Hollow

Firn

Snow formation

Icefall

Convergence

Sediment carried by process of erosion

Snout

Rocks carried by erosion

During glaciation, ice conceals the action of a glacier on the land beneath. But after the glacier melts, the full effects of glaciation become apparent. The most distinctive feature of mountain glaciation is the "U-shaped" valley, which can be found in many hilly and mountainous regions of the world.

BEFORE

A river running down a valley normally cuts a V-shape with sloping sides. Viewed from above, the valley often has a zig-zag shape with the river flowing around rocks projecting from the mountain. When a glacier forms, the ice fills most of the valley and buries the rocks. The valley also appears much straighter.

At the lower end of the valley glacier, the snout (end) of the glacier tapers toward the ground as the ice begins to melt. Along each side of the glacier, a pile of rock fragments collect. These piles of rock fragments are known as a lateral moraine. When the glacier flows over bumps in the bedrock or around a bed, deep, V-shaped clefts may form. These clefts are known as crevasses. Crevasses can be up to 100 feet deep, big enough to swallow a house.

The U-shaped valley created by a glacier moving downhill in Wales

▽ As glaciers flow down a valley, a pile of rocks collects at the sides. These piles are known as lateral moraines.

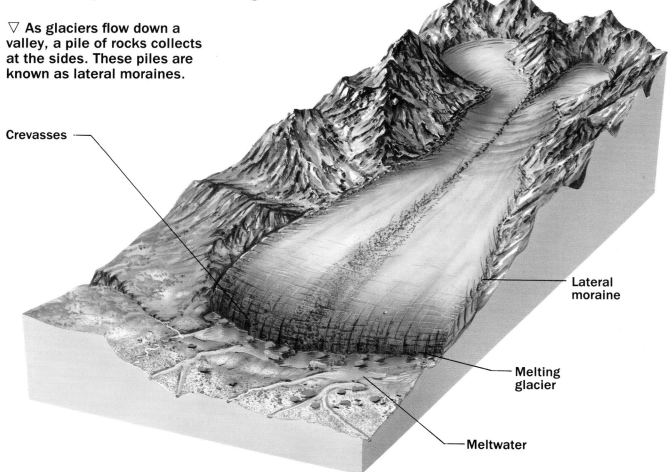

Crevasses

Lateral moraine

Melting glacier

Meltwater

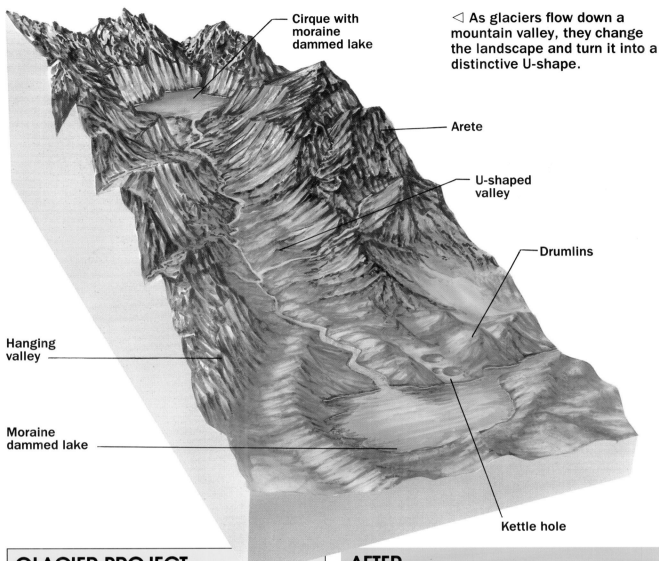

Cirque with moraine dammed lake

◁ As glaciers flow down a mountain valley, they change the landscape and turn it into a distinctive U-shape.

Arete

U-shaped valley

Drumlins

Hanging valley

Moraine dammed lake

Kettle hole

GLACIER PROJECT

Make a U-shaped valley in a sand tray. Firmly push an ice cube (glacier) down the valley. What shape is the valley now?

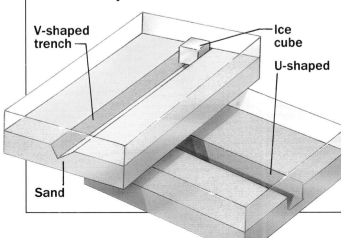

V-shaped trench

Ice cube

U-shaped

Sand

AFTER

As the glacier moves down the valley, it erodes the valley floor and cuts through the rocks. When the glacier melts, the valley looks much deeper and has much steeper sides. The glacier has created its distinctive U-shaped valley.

There are few features visible on the valley floor. The most dramatic signs are along the sides of the valley. The jutting rocks are all cut off and between them, tributary valleys enter the main one half-way up the sides. These hanging valleys are often marked by a waterfall. At the very end of the valley, a lake may have formed behind a terminal moraine, the pile of rocks and debris left behind by the glacier.

Apart from erosion, glaciers also alter the landscape through transport and deposition of rocks. Glaciers contain frozen water and thousands of tons of rock fragments plucked from the valley sides and floor. These fragments are transported by the ice, and are deposited when the glacier finally melts.

TRANSPORT

The rock fragments carried by a glacier vary in size from large boulders to tiny particles of sand. The smaller fragments are collectively known as glacial drift.

Although the fragments are spread throughout the glacier, they tend to be concentrated in the narrow bands of moraines. When two glaciers converge, their lateral moraines combine into a single moraine which runs down the center of the glacier. This is known as a medial moraine.

Dark rock fragments on the surface of a glacier are able to absorb more sunlight than the surrounding white ice. This makes the rocks warm up much faster, and the rate at which the ice melts increases.

Glacier and moraine in the Himalayas

Drumlin

Bedrock

Terminal moraine

Medial moraine

△ Glaciers create many different landforms. The drumlins in the picture are in a field in Britain.

LANDFORMS

The most distinctive feature of glacial deposition is the terminal moraine. If the glacier remains at a standstill over a long period, the moraine will build up into an impressive height. If the glacier melts in stages, a series of terminal moraines may occur across the whole valley.

Drumlins are clusters of low, egg-shaped hills that form behind the snout of a glacier. As a glacier slows down at the snout, it loses much of its power. Instead of smoothening out irregularities in the bedrock, the glacier deposits glacial till around these bumps, producing rounded hills. The shape of the hill indicates the direction of the ice flow, and the blunt end of the drumlin points toward the glacier's snout.

ERRATICS

An erratic is a pebble or boulder that is made of a completely different type of rock from those found nearby. Many erratics were carried to their present location during the last Ice Age. Instead of being turned into glacial drift, erratics were transported whole and frozen inside the ice. When the glaciers melted, the erratics were deposited many miles away from their original locations.

Erratics provide valuable clues to the movement of glaciers during the Ice Age. By finding the source rock (the original location) of the erratics, scientists can trace the path of the glacier on a map. For example, some erratics found in Poland have been traced back to Norway. Others have been traced to a different source in Finland. These results indicate that two huge glaciers must have moved into Poland before flowing down either side of the Baltic Sea.

▽ The map shows the location of some erratics. The picture above shows an erratic resting on a pavement near Black Head in Ireland.

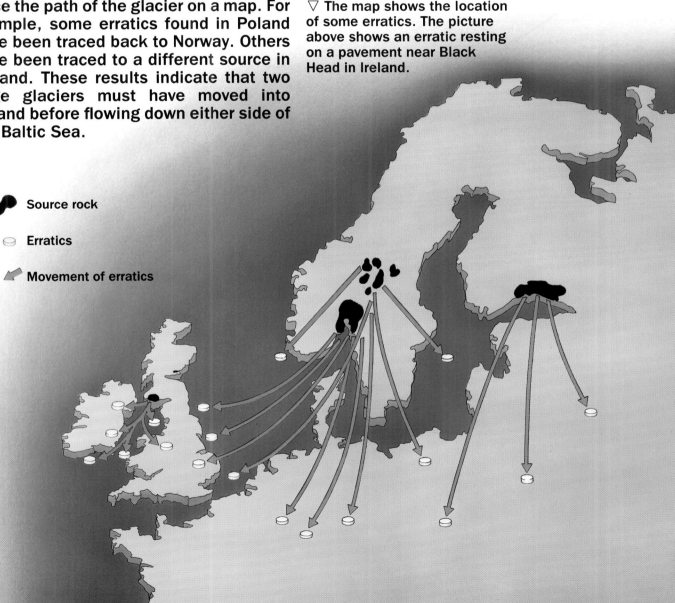

● Source rock

◒ Erratics

◤ Movement of erratics

Beyond a glacier's snout, meltwater continues the process of transport and deposition. Streams and rivers fed by glaciers carry glacial drift far beyond the limits of the ice. Wind may then extend the effects of glaciation on the landscape. Glaciers are also indirectly responsible for sand dunes.

DEPOSITION

The most extensive surface feature created by meltwater is the outwash plain. As temperatures rise and fall during the summer and winter months, streams form within the glacier which sort and redistribute fluvioglacial deposits, creating an outwash plain. An outwash plain is a flat area in front of a glacier composed of rocks and fluvioglacial sediments. In lowland areas, the plains can be tens of miles wide.

The continuous flow of meltwater causes deposits in an outwash plain to become graded according to size. The smallest fragments are transported over the greatest distance. Close to the glacier snout the plain consists of gravel beds. Further away, the gravel in the outwash plain turns to coarse sand.

As the outwash plains dried up at the end of the last Ice Age, huge amounts of powdered rock were carried away by the wind. Deposits of this wind-blown material (called loess) are found in many parts of Europe and the United States.

▽ Glacial debris in Scotland

▷ Glacier meltwater creates many types of landforms, including streams, outwash plains, and steep hills which are known as kame terraces.

Kame

Sediment deposits (smaller further away)

Glacial stream

RIDGES AND MOUNDS

Around the snout of a glacier, the strong flow of meltwater creates a series of distinctive features. For example, meltwater running down the sides of a glacier deposits drift along the valley sides. This drift builds up into rounded terraces that are known as kame terraces.

Streams of meltwater running beneath a glacier form long, low ridges of sand and gravel known as eskers. If the rate of melting increases so does the rate of deposition, and the eskers grow in size. Several periods of melting create eskers that vary in height and width. Eskers also vary greatly in length, and may wind for hundreds of feet across the landscape. The sand and gravel from eskers, kames, and outwash plains are used in the construction industry in Europe and North America.

A glacier behind a volcano in Argentina.

DID YOU KNOW?

The Sand Hills of Nebraska cover more than 19,000 square miles. The sand in these rolling dunes was produced by glaciation and transported to Nebraska by the wind thousands of years ago.

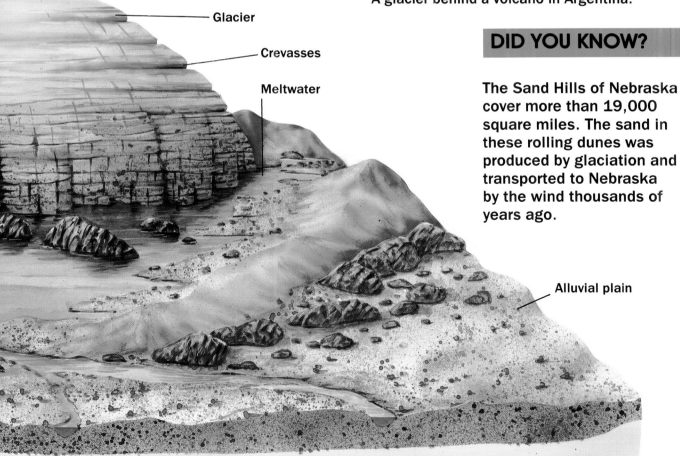

Glacier

Crevasses

Meltwater

Alluvial plain

When the glaciers melted after the last Ice Age, most of the water flowed back into the oceans. However, some of it remained behind in glacial lakes. There are thousands of such lakes scattered across areas that were once glaciated. The most famous of these are the Great Lakes.

FORMATION

Many glacial lakes, including the Great Lakes, formed when meltwater flowed into the natural hollows and basins of the landscape. Other lakes formed on the floors of glaciated valleys, which were much deeper than before. Sometimes the lakes were held back by a moraine, which acted like a natural dam.

Along the coasts, water flowing back to the sea flooded glaciated valleys, creating deep fjords. Norway has many fjords along the coast, the longest of which is Sognefjord at 125 miles. Another kind of lake created by melting ice were the small kettle lakes. These form when bits of ice buried in glacial drift melt, filling up the resulting hollow.

▷ Fjords occur along the coasts. Fjords are often surrounded by cliffs that are over 655 feet high.

Moraine in Wester Ross, Scotland

Valley

Fjord

Basin eroded by glacier

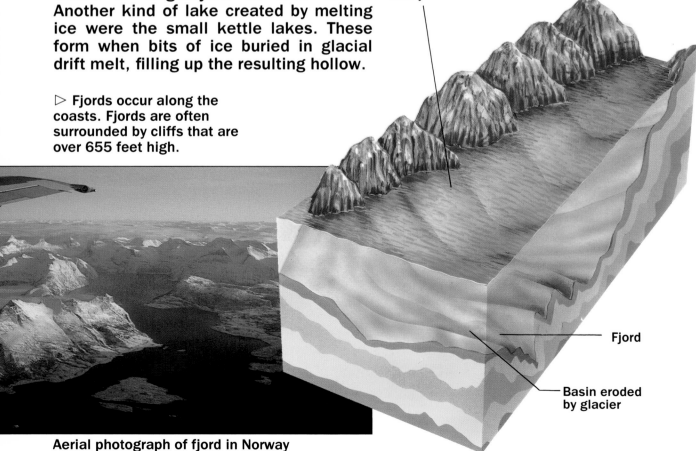

Aerial photograph of fjord in Norway

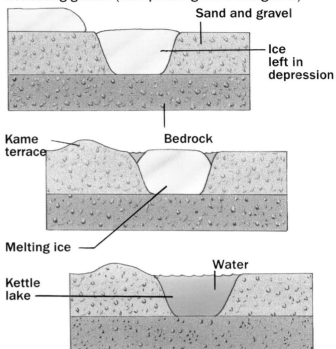

Retreating glacier (incorporating sand and gravel)

Sand and gravel

Ice left in depression

Kame terrace

Bedrock

Melting ice

Kettle lake

Water

△ Kettle lakes form when buried ice melts in glacial deposits. They are often small and circular.

MORAINE LAKES

Moraine lakes reveal a great deal about the behavior of glaciers during the Ice Age. As glaciers melted, lakes often formed between the snout and terminal moraine, the pile of rock debris at the end of the glacier. As more of the ice melted, glacial drift often collected on the bottom of the lake, forming a shallow, cross-valley moraine.

If the rate of melting suddenly increased, the lake could overflow the terminal moraine or the valley sides. When this occurred, the meltwater cut an overflow channel that can still be seen on the landscape today.

Where the meltwater entered a lake from glacier-fed streams, drift was laid down in layers known as varves. Because the glaciers melt every year in spring, the varves occur in alternate layers of silt and sand. Each pair of layers represents one year's worth of melting. By studying varves in the ice, scientists are able to calculate how quickly a glacier melted.

MAKE A KETTLE LAKE

Place an ice block in a bowl and surround it with damp sand. Leave the bowl standing in a warm room for an hour. As the ice melts, water will be retained in the resulting hollow.

Ice

Sand

Bowl

Melted ice in hollow

Ice Ages are caused by long-term variations in the earth's climate. These variations probably occur as a result of changes in the earth's orbit around the sun. Dust clouds from volcanoes may also play some part, for they can block out the sun's rays. Scientists also feel that pollution affects the earth's climate.

TEMPERATURE CHANGE

The key factor affecting temperatures on the earth is the amount of sunlight reaching the planet. A small change in the tilt of the earth's axis would lead to a change in the amount of sunlight striking the earth. Variations in the average distance between the earth and sun would also have the same effect.

The earth's climate is also subject to short-term variations, and the average temperature is constantly rising or falling. Between the 1400s and 1700s, the world experienced the Little Ice Age. Average temperatures fell by about 4°F, and European glaciers began to advance. Between 1880 and 1940 temperatures rose by about 2°F. The reasons for these short-term variations in temperatures are still unclear, and scientists are not sure of present trends.

Ski resorts suffer from fluctuating temperatures.

▽ A flooded region in Bangladesh. Scientists feel glaciers might be be able to affect the world's temperatures.

SEA LEVEL

The earth's climate is delicately balanced. The amount of water on the planet remains constant, whether it is in the oceans, in glaciers, or in the atmosphere. When glaciers were at their greatest extent, sea levels were 330 feet lower than they are today. If the polar ice sheets were to melt away, sea levels would rise to some 230 feet above today's levels.

At present, sea levels are rising at about ¾ to 1 inch every ten years. Such a slow rate of increase presents little threat to human populations. However, if temperatures rose dramatically and the ice sheets began to melt, sea levels would increase much more rapidly. Millions of people who live along the coasts would be threatened by flooding. Many scientists think that pollution could cause temperatures, and sea levels, to rise in the future.

POLLUTION

The earth's atmosphere acts like a giant greenhouse, trapping the sun's heat near the planet's surface. High in the atmosphere, a layer of gases, mainly carbon dioxide, reflect heat back to the earth. These are known as greenhouse gases, and under natural conditions the level of these gases remains fairly constant.

However, over the last century, pollution from activities on Earth has increased the amount of greenhouse gases in the atmosphere. Burning hydrocarbon fuels, such as coal, oil, and gas, releases huge amounts of carbon dioxide. Other industrial gases, such as the CFCs used in refrigerators and aerosols, are extremely powerful greenhouse gases which also escape into the atmosphere. Another reason for the increase in carbon dioxide is the destruction of forests that usually turn this gas into oxygen. Scientists feel that this increase in greenhouse gases might lead to higher temperatures and global warming.

Millions of tons of pollution enter the atmosphere every year.

▽ Pollution from factories and power plants is increasing the amount of greenhouse gases in the atmosphere. Many scientists believe that the earth will get warmer as a result.

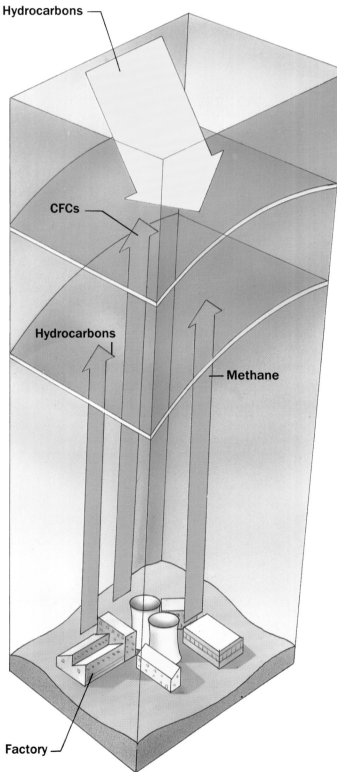

Hydrocarbons

CFCs

Hydrocarbons

Methane

Factory

Millions of years of snowfall have been compressed into the polar ice sheets. During its journey to the ground, each snowflake absorbs minute quantities of gas and dust from the atmosphere. By studying the ice sheets, scientists have made many discoveries about the earth's past climate and history.

DIFFERENT METHODS

During the first half of this century, the thickness of glacial ice was measured by echolocation. Explosions were set off, and the echoes from the underlying rocks were recorded. More recently, radar has been used to measure the thickness to within a few inches. Transmitted from airplanes or satellites, radar waves easily penetrate the ice and are reflected by the rock beneath. Using radar images, accurate maps of the Antarctic continent have been drawn.

Detailed information about the past can also be obtained by taking ice cores from the glacier depths. Scientists obtained these cores by drilling, and cores from depths of 5,000 feet have already been obtained. Such cores contain ice

A scientist researching glaciers

more than 120,000 years old.

Information about more recent years is obtained from ice nearer the surface. Ice from the base of a 100-foot crevasse would tell scientists about how the earth was 2,500 years ago. Occasionally, the ice cores or bases contain large objects. For example, in 1991, an icy region in the Alps between Austria and Italy had a 4,000-year-old human body inside it.

▽ Descending a crevasse is very dangerous.

Most ice stations are located in remote polar areas, like this one in Antarctica.

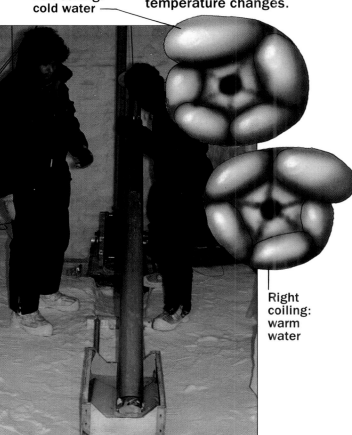

▽ Ice cores provide evidence of climatic change. Minute animals alter the way their shells coil when the temperature changes.

Left coiling: cold water

Right coiling: warm water

THROUGH TIME

Snowflakes contain samples of the earth's atmosphere, and each year's snowfall is a time capsule of atmospheric conditions. When ice cores are taken to the laboratory, the first task is to separate the tightly compressed layers of ice. Each layer is then subjected to chemical analysis, releasing the various gases and solid material dissolved in it. These are then identified and quantified, before being compared with samples from different depths.

Among the first things scientists discovered was evidence of atmospheric pollution. Cores from Greenland reveal that levels of lead began to increase steadily around 1800 with the advent of industrialization. Lead levels shot up after the introduction of leaded gasoline.

Particles from volcanic eruptions can also be detected in the ice. The Greenland ice cores have been used to determine the exact date that Thera, a volcano on the Mediterranean island of Santorini, erupted. Evidence shows that Thera erupted around 1627 B.C., about 100 years earlier than everyone thought.

Antarctica is a unique place — a vast and virtually untouched wilderness of ice and snow. So far, the Antarctic wilderness has been preserved through scientific cooperation between countries. However, increasing world demand for raw materials means that the future of Antarctica is by no means certain.

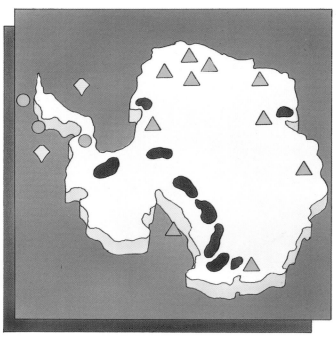

Frozen seal in Captain Scott's hut in Antarctica

◇ Gold   △ Metals   ○

No country owns Antarctica, although several of them have laid claim to parts of it over the years. Nor is Antarctica an independent state. At present, the region is administered by an international treaty which prohibits most forms of development and mining.

International cooperation in Antarctica began with scientific expeditions in the late 1950s. At that time (during the Cold War between the United States and the Soviet Union), there was considerable fear that Antarctica might be used for military purposes. The Antarctic Treaty came into force in 1961 to ensure that it was used for peaceful purposes.

Recently, fears about pollution have brought Antarctica into the headlines again. The reason for this is that once pollution reaches Antarctica, it is likely to stay there forever. The Antarctic climate will not disperse pollution, but will preserve it. The pollution also threatens the highly productive ecology around the Southern Ocean.

Beneath the ice and snow in Antarctica, valuable deposits of gold, iron ore, and other metals have been discovered. As a result, some countries are now reluctant to ban mining and other forms of mineral extraction from Antarctica.

The possibility of large-scale mining operations in Antarctica has alarmed many of the world's environmentalists. They argue that Antarctica should be declared an "earth park," free from any form of development or mining.

A British scientist measuring snow density in Antarctica.

Cirque
A bowl-shaped depression on a mountain caused by erosion at the head of an alpine glacier.

Continental glacier
An ice sheet that is a large part of a continent, for example, the Antarctic ice sheet.

Drumlins
Rounded hills formed by a deposition from a glacier.

Erratic
Rock fragment or boulder transported far from its source.

Esker
Long mound of drift deposited by a meltwater stream under a glacier.

Firn
Halfway stage in the transformation of snow into glacier ice.

Ice Age
Period when the earth's climate was colder that today; the glacial period.

Ice cap
Ice sheet; lens-shaped glacier up to 19,000 square miles in area.

Ice sheet
Often large, lens-shaped glacier that completely covers the underlying landscape; ice cap.

Ice shelf
Thick layer of ice that extends out to sea.

Kame Terrace
A low, steep-sided hill of stratified drift formed in direct contact with glacial ice.

Kettle lake
Small lake formed by melting block of glacial ice.

Lateral moraine
Mound of rock fragments carried at sides of glacier.

Loess
Sand and soil that has been deposited by the wind.

Moraine
A concentration of rock fragments transported by a glacier.

Outwash plain
Flat area of graded material formed ahead of melting glacier.

Pack ice
The frozen surface of the sea.

Permafrost
Ground that is permanently frozen except for a thin surface layer.

Pingo
Crater formed by ice block erupting onto surface.

Snout
Down-valley end of glacier; ice front.

Terminal moraine
Fragments formed at snout of glacier.

U-shaped valley
Valley that has been widened and eroded by a glacier.

Valley glacier
Glacier enclosed in a mountain valley.

Varve
Seasonal layers of drift down in a moraine lake by glacial meltwater.

Photographic Credits:
Cover and pages 14, 16 & 22: Planet Earth Pictures; introduction page & pages 9, 27 & 28 left: Frank Spooner Pictures; 6, 10 bottom & 13: Spectrum Colour Library; 10 top, 11, 12, 14-15, 20 bottom, 21, 24-25 all, 28 right, 29 top & bottom & 30 bottom: Science Photo Library; 15, 20 top & 23: Bruce Coleman Ltd; 18: NHPA; 26 top: Topham Picture Source; 26 bottom: Roger Vlitos; 30 top: Frank Lane Picture Agency.